essentials
Springer essentials

T0074060

Springer essentials provide up-to-date knowledge in a concentrated form. They aim to deliver the essence of what counts as "state-of-the-art" in the current academic discussion or in practice. With their quick, uncomplicated and comprehensible information, essentials provide:

- an introduction to a current issue within your field of expertise
- an introduction to a new topic of interest
- an insight, in order to be able to join in the discussion on a particular topic

Available in electronic and printed format, the books present expert knowledge from Springer specialist authors in a compact form. They are particularly suitable for use as eBooks on tablet PCs, eBook readers and smartphones. *Springer essentials* form modules of knowledge from the areas economics, social sciences and humanities, technology and natural sciences, as well as from medicine, psychology and health professions, written by renowned Springer-authors across many disciplines.

Friedrich Frischknecht

Malaria

Deadly parasites, exciting research but no vaccination

 Springer

Friedrich Frischknecht
Department für Infektiologie, Abteilung Parasitologie
Universitätsklinikum Heidelberg
Heidelberg, Germany

ISSN 2197-6708 ISSN 2197-6716 (electronic)
essentials
ISBN 978-3-658-38406-7 ISBN 978-3-658-38407-4 (eBook)
https://doi.org/10.1007/978-3-658-38407-4

This Springer Spektrum imprint is published by the registered company Springer Fachmedien Wiesbaden GmbH, part of Springer Nature.
The registered company address is: Abraham-Lincoln-Str. 46, 65189 Wiesbaden, Germany

Preface

Malaria is an infectious disease caused by parasites that has accompanied mankind like no other since its origins. In the twentieth century, 200 million people died of malaria, and today, 'only' 400,000 per year, mostly children under 5 years of age, after they were infected with *Plasmodium falciparum* for the first time through the bite of an *Anopheles* mosquito. If the children survive the first infections, they slowly build up their own protection against malaria. This gives rise to the hope that a vaccination against malaria will one day be possible. However, this hope has repeatedly met with disappointment in recent decades despite intensive research. With the present, very short *essential,* I would like to present the basics of the way of life of the malaria pathogens to the interested public and students of all disciplines. The aim is also to achieve an understanding of the complex relationships between the parasite, humans, and the environment. While malaria is a tropical disease today, it used to be widespread far into the cold north. There it was only defeated in the 1970s. But will we be able to eradicate it in the tropics?

I would also like to take this opportunity to say thank Markus Ganter for discussions and Sarah Koch for motivating me to take on this DeepL-driven translation of the text from its German original.

Heidelberg, Germany   Friedrich Frischknecht

What You Can Find in This *essential*

- The complex life cycle of the malaria parasites, simply explained.
- When does malaria make you ill and how often?
- Is malaria a disease of the tropics or of poverty?
- How are parasites examined?
- How are vaccines developed?
- Could we eradicate malaria or is it coming back to Europe and the USA?

Contents

Friedrich Frischknecht studied biochemistry at the Free University of Berlin and did his doctorate on pox viruses at the European Molecular Biology Laboratory (EMBL) in Heidelberg. Following a research stay at Institut Pasteur in Paris, he has been head of a research group at the University Hospital in Heidelberg since 2005 and is working on the molecular basis of the movement of malaria parasites. He has received several awards for his research work. Frischknecht collaborates with colleagues from different countries and regularly teaches in Africa. For more information, short video clips about his current research and teaching, and comic books about malaria: www.sporozoite.org.

Friedrich Frischknecht, Prof. Dr., University Hospital Heidelberg, Im Neuenheimer Feld 344, 69120 Heidelberg. Photo by Yves Sucksdorff

Introduction

1

Abstract

Malaria is probably the most important infectious disease in the history of mankind and unfortunately its eradication is not foreseeable. Why this is and what the challenges towards eradication of malaria are shall be presented in this *essential*.

Malaria has left traces in humans and their history like no other infectious disease. Famous commanders and rulers suffered from malaria as well as millions of their subjects. Armies were destroyed and wars were decided by the disease. Unlike the epidemics of smallpox or the plague, which led to the decline of many cultures, malaria was present every year. Only in cold winters or dry summers, when the mosquitoes that transmit malaria could not lay eggs, did it give people a short break to breathe. With the discovery of the cinchona bark as a remedy against malaria, man began his triumphal march againt this age-old scourge. This march is still accompanied by successes such as the eradication of malaria in China in 2017 but also by serious setbacks. For example, in 2016 the number of malaria cases worldwide increased again for the first time in the twenty-first century and a much awaited first vaccine is not very efficient. While the last person fell ill with smallpox in 1977 and the World Health Organization publicly announced the greatest achievement of mankind, namely the eradication of the smallpox virus, in 1980, malaria could only be banned from parts of the globe to this day. With this booklet, I would like to present the variety of tiny malaria parasites to interested laypeople. I also want to show how scientists are trying to produce an efficient vaccine against malaria and why this effort has so far not been successful.

© Springer Fachmedien Wiesbaden GmbH, part of Springer Nature 2023
F. Frischknecht, *Malaria*, essentials,
https://doi.org/10.1007/978-3-658-38407-4_1

From Swamp Fever to Plasmodium

2

Abstract

Malaria, or swamp fever, is a disease of poverty that has accompanied mankind since the beginning of civilization. However, the pathogen and the way it spreads via mosquitoes were not discovered until the late nineteenth century. Today malaria is limited to tropical areas.

The term malaria is composed from the Italian words for bad (mal) and air (aria), and indicates the long known connection of the occurrence of the disease with the stagnant, sultry, and bad air in and around swamps. This is why malaria was also called marsh fever for centuries. While miasma, bad air, were once seen as the cause of the disease, the pathogens known today as causing malaria are various types of unicellular parasites of the genus *Plasmodium,* which are transmitted by mosquitoes.

2.1   Symptoms of Malaria

After the bite of a mosquito, depending on the parasite transmitted, between seven and 30 days pass before the first symptoms appear. These can be light or high fever, tiredness, headache, chills, heavy sweating, malaise, vomiting, and general body aches. The symptoms are little different from those of influenza and the often quoted classic picture of periodic fever attacks is not always observed. Malaria is therefore often detected late in countries where the disease does not occur. In countries where malaria is widespread, on the other hand, bacterial or viral diseases are often overlooked and not treated because no alternative causes for the symptoms are sought. It must therefore be assumed that many deaths attributed to malaria in

© Springer Fachmedien Wiesbaden GmbH, part of Springer Nature 2023   3
F. Frischknecht, *Malaria*, essentials,
https://doi.org/10.1007/978-3-658-38407-4_2

these countries are due to other pathogens. A clear diagnosis can be life saving and is therefore essential. In well equipped laboratories, this can be done reliably by simple procedures, but if poorly equipped, for example in developing countries, it can lead to false diagnoses. And to add even more complexity, the sick person can harbor parasites showing a malaria infection, yet the symptoms can come from other pathogens as many people who had malaria many times can control parasite growth to a degree where the parasites are still present but do not cause disease.

2.2   Malaria on the Rhine

Malaria is now largely confined to countries in the tropics and is therefore referred to as a tropical disease (Fig. 2.1).

However, this was not always the case. Only in the 1970s was malaria driven out of Europe and the USA. Even in the coldest years of the Little Ice Age around 1600, malaria occurred in England and was spread as far as the north of Siberia. An impressive testimony of malaria and the burden it imposed on the people is given in a section of a letter from the German poet Friedrich Schiller to Caroline von Wolzogen from September 11, 1783:

> I've been ill for three weeks, my Dearest. Without danger to my life, thank God, but a cold fever, of which I had to endure a daily attack, has taken a terrible toll on me, and although I have already recovered except for the weakness of my head, I will not leave the house for another fortnight. In the eight weeks I spent in Mannheim, a gale-like epidemic is raging in the city, which is so widespread that 6,000 of the 20,000 people in the city are ill. Meier died of it while I was here. A friend I owed a lot to. Now – thank God – the epidemic is on the decline.

Fig. 2.1 Morning fish hunt in a stagnant water body at the Sepik River in Papua New Guinea People along the river regularly contract malaria. (Source: Recorded by the author in 2012)

1783 was the year, the United States of America were recognized as a souvereign country and Mannheim, where the Neckar river flows into the Rhine became the birthplace of both the bicycle, in 1817 and the automobile, in 1885. George Washington, the first president of the United States suffered from malaria, as he did from smallpox and tuberculosis. Malaria is therefore not a tropical disease per se, but a disease of the swamps. It is also a disease of poverty: Poverty of knowledge and poverty of financial possibilities. Malaria was driven out of certain areas even without the knowledge of what caused malaria and how the disease was transmitted from person to person. Thus, when the Rhine between Basel in Switzerland and Bingen in Germany, some 300 km to the North, was 'regulated' in the nineteenth century, it largely disappeared from the Rhine valley (Fig. 2.2). The so called regulation of the Rhine is paraphrasing one of the most gigantic destructions of nature that mankind has ever performed; it destroyed more than 2000 islands in the Rhine

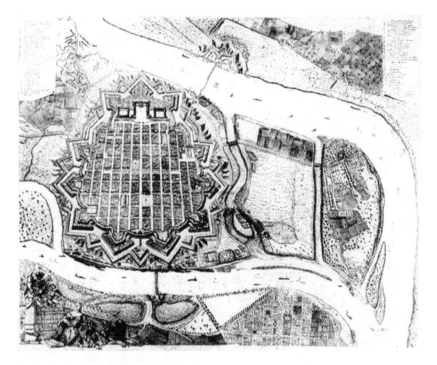

Fig. 2.2   Mannheim in the eighteenth century – *above* the Rhine, *below* the Neckar, note the moats around the fortress walls. The city now has over 300.000 inhabitants, several universities and a thriving industrial and cultural sector. (Source: Wikipedia (public domain) – copper engraving by Josef Anton Baertels 1758)

river ecosystem. With the improved water discharge created by the shifted river flow, there were fewer swamps on the one hand and increasing trade on the other. As a result, the people along the river became more prosperous and were able to build better houses without the fear of losing them to flooding. The more sturdy houses in turn made it more difficult for the mosquitoes to take their blood meals on humans. So malaria decreased and had almost completely disappeared from the Rhine valley before the pathogens (Plasmodium species) and their carriers (*Anopheles* mosquitoes) were discovered. Nevertheless the mosquitoes survived and thrive until this day and are considered a veritable plague by the people living along the Rhine today, even in the absense of transmittable pathogens.

2.3   Discovery of the Pathogens

The microscope was (and still is) the most important instrument in biomedical research. Invented in the early seventeenth century, it became known to a broad public through the sensational observations and drawings of Robert Hooke presented in his book *Micrographia* in 1665. Antonie van Leeuwenhoek, a Dutch merchant who was enthusiastic about *Micrographia,* made his own microscopes and discovered the microcosm of living beings he called *Animalcules.* He saw unicellular organisms of various forms and their movements, described sperm and bacteria, and carried out a first experiment to find out whether the latter can also cause diseases: He examined the microflora in a rotten tooth under his microscope. Many of the small animalculs swam around. He then rinsed his mouth with vinegar and took another sample: Only a few of the bacteria were still active. He concluded that they could cause tooth decay. However, it was not until almost 200 years later that the germ theory of pathogens as the causative agents of disease was proven by Louis Pasteur and Robert Koch and many of their colleagues who discovered the pathogens of bacterial diseases. Because of the long known (see Schiller's descriptions) link between swamps and malaria, it seemed a reasonable hypothesis to search for the pathogen of the disease in swamp water. Two scientists, who had studied under the famous Berlin-based medical doctor and scientist Rudolph Virchow, and who had made many important discoveries themselves, probed this hypothesis. And in 1878 they, Edwin Klebs and Corrado Tommasi-Crudeli, isolated a bacterium from the swamps near Rome and called it *Bacillus malariae* (Cox 2010). Injected into rabbits, it caused a malaria-like fever and caused an enlarged spleen. The riddle seemed to be solved: Malaria is triggered by the ingestion of contaminated water.

Two years later, however, a completely unknown military doctor, Alphonse Laveran, reported that he had discovered a kind of parasite in the fresh blood of 148 of his 192 soldiers suffering from malaria in French-occupied Algeria. After an initial dispute over this new description, it was quickly confirmed by several researchers and doctors and the parasite was eventually named *Plasmodium falciparum* (Cox 2010). Laveran also speculated that mosquitoes could be carriers of the parasite, which could pass it from one person to another through their bites. However, it took another 18 years before this discovery was made by Ronald Ross on malaria in birds. Shortly afterward Battista Grassi showed that in humans only mosquitoes of the genus *Anopheles* transmit malaria, which could explain why in some areas, despite the presence of many mosquitoes, there was no malaria: There were no *Anopheles* mosquitoes (Grassi 1901).

Life Cycle of Malaria-Causing Parasites

3

Abstract

Malaria parasites are spat into the skin by mosquitoes. There they penetrate blood vessels and infect cells in the liver. They multiply thousands of times in liver cells without causing symptoms. Leaking from the liver, the parasites infect red blood cells. The explosive re-infection of red blood cells leads to the disease. The parasites also develop sexual forms that fuse in the stomach of the mosquito after it has ingested blood. The resulting new form of the parasite can break through the stomach wall and infect the mosquito. In cysts on the stomach wall new parasites develop which then penetrate the salivary gland and can be transmitted again.

3.1   Different Species in Humans and Animals

Malaria parasites multiply in the blood and specifically in the red blood cells, which are important for oxygen transport. Curiously, some people will die with less than 0.1% of infected red blood cells, while others will live with up to 40% of their red blood cells being parasitized. The latter corresponds to almost 1 kg of parasite mass that ciruculates. How can this be? Let's first examine the complex live cycle of the parasite before returning to this question later. The pathogens penetrate the blood cell and multiply there. After the parasite has penetrated the red blood cells, it changes its shape so that it resembles a ring for several hours. Then the parasite begins to grow by feeding on the red blood pigment (hemoglobin). It is now called trophozoite, (Greek for 'to eat'). Eventually it multiplies its genetic

© Springer Fachmedien Wiesbaden GmbH, part of Springer Nature 2023
F. Frischknecht, *Malaria*, essentials,
https://doi.org/10.1007/978-3-658-38407-4_3

Female mosquito

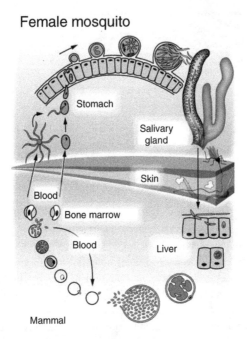

Fig. 3.1 Life cycle of *Plasmodium* – one mosquito spits 10–100 parasites into the skin. With the flow of blood they enter the liver where they invade and multiply in liver cells. Just 1–10 parasites develop in the liver into tens of thousands of parasites. These are released into the bloodstream where they infect red blood cells to multiply. The sexual forms develop in the bone marrow and can be taken up by blood feeding mosquitoes. Fertilization takes place in the mosquito stomach. 10–100 new parasites develop, which penetrate the stomach wall and form 1–10 cysts. In these cysts up to 5000 new parasites are formed. These penetrate the salivary gland and are spat back into the skin with the next bite. (Source: Yves Michel Cully and Friedrich Frischknecht)

information and forms new parasites, which then leave the cell explosively to attack new red blood cells (Fig. 3.1).

The reproduction of parasites takes place in an unusual way: Normally, cells duplicate their genetic information, which is present in cell nuclei, during cell division; however, malaria-causing parasites form a growing cell, while they duplicate their genetic information and their cell nuclei several times within the cell. After a certain number of cell nuclei have formed, new parasites form around the individual nuclei. This process is superficially similar to the multiplication of viruses, although the parasites are much more complicated. They have about 5500 genes which compares to around 250 genes of poxviruses or just 10 of HIV. Malaria parasites also contain different organelles, such as the cell nucleus and a chloroplast-like organelle that has changed with evolution. This chloroplast, however, cannot convert light into energy, as in plants, which makes sense as the parasites replicate essentially in darkness. But nevertheless this chloroplast-like organelle is important for the synthesis of fatty acids. It is therefore an interesting target for new active substances that inhibit biochemical processes in this organelle and are explored and used to kill the parasites.

People can be infected by different types of malaria parasites. The already mentioned *Plasmodium falciparum* triggers the most devastating form of disease and is the most important pathogen. It causes the most infections and deaths worldwide. In 2017 it was responsible for 98% of deaths of malaria patients. The medically second most important parasite is *Plasmodium vivax*. It occurs mainly in Asia and South America. *Plasmodium ovale* and *Plasmodium malariae* can also infect humans. These occur in Africa. In addition, there are at least two other Plasmodium species that actually only infect monkeys, which can also cause malaria in humans: *Plasmodium knowlesi* (Asia) and *Plasmodium simium* (Brazil) (Moon et al. 2013; Brasil et al. 2017). These can be fatal, especially in the case of infections with *Plasmodium knowlesi*.

The different species can be distinguished to a large extent with the aid of a simple microscope. A drop of blood is spread out on a microscope slide, fixed in place and then stained in a solution named after the researcher Gustav Giemsa. More precisely, this solution stains the parasites with a different hue than the red cells and thus enables not only their detection, but also the differentiation of the different species. In *P. falciparum,* only the so-called rings are visible in the stained drop of blood, whereas in *P. vivax* the later stages (e.g., trophozoites) are also visible. In *P. falciparum* only the rings are visible, as the red blood cell changes during the course of an infection in such a way that it attaches itself to the wall of thin blood vessels in various organs, so that only the early stages circulate in the blood stream. *P. falciparum* may have developed this change in its host cell during evolution to avoid the red blood cell infected by the parasite being 'discarded' by the spleen. The spleen controls the 'health' of the red blood cells and, for example, takes old red blood cells out of circulation. This succeeds because the stiffness of the cells increases with age and they then get stuck in the spleen, where they are immediately absorbed by scavenger cells, so-called macrophages. Since infection with *P. falciparum* also leads to stiffening of the red blood cells, these cells would be discarded in the same way unless they adhere to the wall of the blood vessel. In contrast to *P. falciparum, P. vivax* prefers to infect young blood cells and does not stiffen them. So there was no need for *P. vivax* to develop a system to adhere to the blood vessel walls and all forms of the parasite circulate in the blood.

In addition to the parasites that infect humans and monkeys, there are also a variety of malaria parasites (almost 200 described species) that can infect birds, lizards, and mice, among others. As mentioned above, research on bird malaria has played a crucial role in the discovery of the life cycle of malaria parasites. Research on bird malaria parasites is still going on today as they allow valuable insights into host-parasite interactions and occasionally there are reports of the death of, for example, penguins in zoos by malaria. The parasites causing bird malaria are found throughout Europe and sometimes infect up to 30% of songbirds without killing them. In the laboratory, malaria pathogens are often used which multiply in mice and rats and which can be studied more closely by means of animal experiments, which would not be feasible with human malaria parasites. Such animal experiments have been and still are crucial in the discovery of bio-active compounds that kill parasites and the development of new vaccination strategies.

3.2   The Path Through the Mosquito

Female mosquitoes bite to drink blood and ingest the parasites. In order to develop in the mosquito, however, it is not enough for the parasite to simply multiply in the red blood cells: The parasite must already transform into the so-called gametocytes in the human blood. These are specialized 'male' and 'female' cells from which a male or female germ cell can develop in the mosquito's stomach (Fig. 3.1). Gametocytes of *P. falciparum* develop over different stages. The youngest ones 'migrate' into the human bone marrow and stiffen there as they develop over days. The stiffening serves to prevent the parasites from re-entering the bloodstream for the time being. Eventually, they become soft again and are thus returned to the bloodstream. Only then can the now infectious gametocytes be absorbed by the mosquito. Surprisingly, Viagra is able to maintain the stiffening of the gametocytes and prevent them from leaking out of the bone marrow and thus the transmission of the parasites to the mosquitoes (Ramdani et al. 2015). While the similarities between human erectile dysfunction and effective malaria transmission are being biochemically investigated, cell stiffening is measured with atomic force microscopes. These discoveries therefore also show the interesting facets of modern malaria research, combining expertise from biochemistry and physics with infection biology and medicine.

After the gametocytes have arrived in the mosquito's stomach, suddenly everything happens very quickly: A program of cell differentitation is activated and produces eight male germ cells which, in search of a female germ cell (egg), swim like sperm through the contents of the insect's blood meal. The female germ cells also release themselves from the red blood cell in order to be fertilized. A colleague once paraphrased that malaria parasites must evidently also undress in order to produce offspring. The fusion of the male and female germ cell leads to the mixing of the genes in the fertilized egg. This leads to the development of a wide variety of pathogen strains, especially in areas with a high rate of malaria infection. Among other things, this high diversity prevents the development of an effective vaccine (Matuschewski 2017).

The fertilized egg cell now develops into a mobile egg and is called the ookinete. This ookinete drills its way from the stomach contents through the stomach wall. The parasite stops its migration on the other side of the stomach wall and forms an oocyst (Fig. 3.1). In this parasitic cyst, the next form of the parasite, the so-called sporozoites, develops (Fig. 3.2). For this purpose the parasite grows in the oocyst and multiplies its genetic information. The cell organelles grow and divide. In the course of several days to 2 weeks, hundreds to thousands of cell nuclei and

Fig. 3.2 *Plasmodium* sporozoite – this form of the parasite is formed in cysts on the stomach wall of the mosquito and penetrates the salivary glands. When the mosquito bites, sporozoites are spat into the skin. From there the parasite migrates into the bloodstream and infects the liver. The parasite is a hundredth of a millimeter long and always appears slightly curved. (Source: Scanning electron microscopy courtesy of Dr. Leandro Lemgruber, University Hospital Heidelberg)

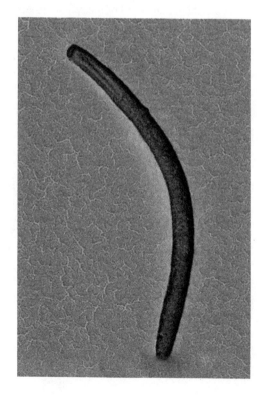

organelles are generated, and eventually form as many sporozoites. The fully formed sporozoites move slowly and emerge from the oocyst. They have now arrived in the body fluid of the mosquito and are driven around the entire body by the insect's rapid heartbeat. However, they specifically attach themselves only to the insect's salivary glands and penetrate them. There they wait in the salivary cavities or salivary ducts until the mosquito transmits them with the next bite. In the process, about 10–100 sporozoites are spat into the human skin. This transmission of the parasites happens before the insect sucks the blood. The mosquito first pokes around in the skin looking for the blood vessel from which it can drink. And during this short phase of probing the skin, the parasites are being ejected with the insect's saliva. Even after the mosquito drinks, sporozoites are still spat out. However, these do not make it into the bloodstream, because the suction effect is too strong, so the parasites, instead of being released into the blood flow are flushed back into the mosquito's stomach, where they are digested along with the blood cells. Those sporozoites that have made it into the skin, however, move at a very high speed through the subcutaneous tissue in search of a thin blood vessel into which they can penetrate. About one third of the sprozoites accomplish this feat, while the others remain in the skin or are transported away with the lymph and are killed in the lymph node.

Thus, a single parasite (ookinete) that turns into an oocyst is sufficient to maintain the cycle. Also, a single parasite (sporozoite) that makes it into the bloodstream and out of it into the liver, where it can multiply into thousands of parasites, is enough to reinfect blood cells (Cowman et al. 2016).

3.3   The Liver

The discovery of each phase of the life cycle is always accompanied by a little mystery. The famous researcher Fritz Schaudinn (co-discoverer of the syphilis-causing spirochetes) once reported that the parasites (sporozoites) transmitted by mosquitoes directly attack the red blood cells (Cox 2010). This was long accepted, but several observations cast doubt on this discovery. Interestingly, in the early twentieth century, neurosyphillis was cured by deliberate infection of patients with malaria parasites that caused high fever and hence killed the syphilis bacteria (Freitas et al. 2014). In order to have a reliable source of parasites, volunteers had to be infected with mosquito bites. When blood was taken from these volunteers 30 minutes after the bites and injected into other volunteers, they too became ill. However, if one waited 1, 3 or 5 days to draw blood, no more volunteers could be infected. It seemed as if the parasite had disappeared from the blood and was hiding somewhere.

To solve this mystery and find the hiding place of the parasites, researchers infected a monkey shortly after the Second World War through hundreds of mosquito bites and additional injections of millions of isolated sporozoites (Shortt et al. 1948). Then they waited 7 days and killed the monkey. They made tissue sections and found out that the parasites were only found in the liver. There they multiplied before they were released into the bloodstream. This process is similar to the process in oocysts, except that in the liver cell with about 20,000 parasites, about ten times as many parasites mature. However, not all parasites that have implanted themselves in the liver seem to mature. The liver cell defends itself and kills about half of the parasites before they can fully develop. Only in 2006, researchers in Hamburg and Paris found out that the parasites detach themselves from the infected liver cells in small vesicles and enter the blood in these little bubbels (Sturm et al. 2006). This shows that even today fundamental discoveries can be made about malaria parasites (Wirth and Alonso 2017).

A fascinating and largely unexplained developmental phase of some malaria parasites is the dormant liver phase. *P. vivax* parasites, for example, can enter the liver, start to grow, and while most of them develop and enter the bloodstream, a small number can enter a dormant phase that can last for several years or even decades. For unknown reasons, the dormant parasites in the liver can resume their growth and release parasites into the blood, which in turn cause malaria. This can happen even when the carrier of the dormant parasites is no longer in malarial areas. The resulting disease is often not recognized as malaria, but rather classified as influenza. Therefore it is important to determine the pathogen in a malaria disease. If the patient is infected with *P. vivax*, it is possible that such late 'new infections' can occur without a mosquito bite. Currently there are only two drugs that kill these dormant stages, but not without the risk of side effects (Table 3.1).

Table 3.1 Selected drugs against malaria

Quinine	Still used in some hospitals for severe malaria
Chloroquine	No longer recommended in most countries due to parasite resistance, still good for *P. vivax*
Mefloquine (Lariam)	No longer recommended due to strong side effects
Malarone	As travel prophylaxis
Artemisinin combination therapies	Various preparations – preferred treatment methods
Doxycycline	Also possible as travel prophylaxis
Primaquin and Tafenoquin	Are effective against the liver stages of *P. vivax*

Equally fascinating is the difference in the early development of the parasite in lizards and birds. In these animals, the parasites do not invade the liver at all, but attack cells in the skin and blood vessels where they multiply before entering the blood. Another interesting variation is found in bats, where the parasite develops in the liver. It also invades red blood cells, but immediately forms gametocytes, so that it does not attack new red blood cells at all, but can be directly taken up again by mosquitoes.

3.4   Sickle Cell Anemia and Similar Diseases That Protect Against Malaria

Malaria parasites probably first infected people in western Africa and then spread along with their migrating hosts. The long period of interaction and confrontation between parasites and humans left its traces in our genetic information and today allows us a gruesome insight into the evolution of host and pathogen. The genetic information consists of so-called genes, information units that contain the blueprint for proteins from which our cells are built. Proteins are made up of different sequences of 20 different amino acids that exhibit different biochemical properties. The infinite combinatorial possibilities of arranging these amino acids are the reason for the incredible diversity of life on Earth. However, our genetic make-up, like that of all living things, is not set in stone, but is constantly changing. No human genome is like the next, all are subtly different. Yet, some differences are kept as they confer a clear advantage and can be found in large numbers of people. This has been nicely shown by studies on changes (mutations) in the genetic material that protect against malaria. These mutations can be quite different: One amino acid may be replaced by another, thereby changing the function of a protein. An entire gene can also be lost and with it the blueprint for a particular protein. These changes happen constantly and randomly both in the parasite and in humans and lead to subtle differences in our genetic information. If they do not result in a major disadvantage, they may occur in small numbers in a population of humans. If the change results in a severe disadvantage and the human dies before she can have children, the change is lost.

However, a mutation can also have advantages, for example, it can protect against malaria or against the symptoms of severe malaria. This can then lead to a spread of the mutation, as people can survive a disease with it and have children, perhaps more children than their peers who do not have this change. In this case the mutation spreads in the population and a small piece of evolutionary history has been written.

With the origin of malaria in western Africa, it is not surprising that many such 'useful' mutations in the human genome developed there. The most famous example is the change that leads to sickle cell anemia: A single mutation in the gene that contains the blueprint for beta-globin. Together with alpha-globin, beta-globin forms the important hemoglobin, the red blood pigment. Hemoglobin is the main component of the red blood cells, absorbs oxygen in the lungs and releases carbon dioxide from the tissue there. About 7000 years ago, a change in this gene, a so-called point mutation, occurred. This led to a change in the sixth position of beta-globin. While before all humans had a beta-globin featuring the amino acid glutamic acid at this position now it was replaced by the amino acid valine. These two amino acids have different biochemical properties, for example, glutamic acid is significantly more acidic than valine. This change now led to a small but important difference in the behavior of beta-globin and thus of hemoglobin: Red blood cells infected by *P. falciparum* could no longer attach themselves so well to the wall of the blood vessels and the parasite could no longer reproduce so well in the red blood cells (Archer et al. 2018; Cyrklaff et al. 2016). As a result, the parasite grows more slowly and can be sorted out better in the spleen. Therefore, infected people no longer suffered as much from malaria, and above all, they no longer died from it as frequently. This 'good' mutation was therefore able to establish itself in the human population, and the parasite also gained from this, because it can only reinfect mosquitoes from living humans.

Unfortunately, this story of a small change in genetic information does not only have good sides and happy endings. We humans almost always have two copies of our more than 20,000 genes in our genome, distributed over 46 chromosomes. One copy of a gene and 23 chromosomes are inherited from the mother, the other copy of the gene and the other 23 chromosomes are from the father. Only men are a lillte more complicated having one Y and one X chromosome, which essentially makes us men. However, the beta-globin genes is not located on these sex-specific chromosomes, we hence have two copies of it. If only one copy of the beta-globin in a human being has the sickle cell mutation, then protection against malaria is given without the human being experiencing a restriction to his or her life and well-being. However, if a child inherits the mutation from its father and mother, that is, carries a mutation on both copies of the beta-globin gene, it suffers from sickle cell anemia. Most people only survive this serious disease for a few years and often die before they can have children themselves. Despite better medical care, still thousands of people die from sickle cell disease every year. This shows the bitter coldness of evolution: A mutation will spread if it has an advantage for the population, even if it has a serious disadvantage for some individuals. However, this particular mutation cannot spread to all people and in some parts of West Africa 'only' up to 25% of the population have the mutation.

In contrast to the sickle-cell mutation, some mutations have a 'greater power', such as the one that led to the loss of a protein on the surface of the red blood cells, also in Africa. This has no effect on infection by *P. falciparum,* but *P. vivax* can no longer infect these cells. This mutation, since it does not have any disadvantage, could spread widely and is the reason why *P. vivax is* no longer present in large parts of Africa. However, over time, the parasite seems to have found another way into the cells, which now leads to infection of people in parts of the world who have this mutation and were previously protected from *P. vivax* (Ménard et al. 2010; Mendes et al. 2011). These new discoveries show the continuous evolution between host and pathogen.

Genetic Engineering and Vaccines

4

Abstract

When people repeatedly get malaria for a long time, they develop protection against the parasites and the disease. This observation forms the basis for many experimental vaccines which, unfortunately, have not provided efficient protection against the disease in clinical trials. Attenuated parasites can be used as vaccines. Whether modern genetic engineering can produce better vaccines will only be answered in the coming decades.

4.1 Immunity in Adults

At the end of the nineteenth century, Robert Koch investigated cases of malaria in Papua New Guinea, then 'Kaiser-Wilhelmsland', a German colony. From the plantation village of Stephansort, which has now disappeared in the jungle, he examined the inhabitants and categorized them according to age groups. In one village, he found that eight out of ten examined infants under the age of two had malaria. However, of 12 children aged between 2 and 5 years, only five had malaria, and of the 86 older people examined, none had any. In another village, he found six out of six infants infected, while only ten out of 30 of the 2 to 10-year-olds had malaria. Of 39 older people, none had malaria. From these observations he concluded in 1900 that the inhabitants of tropical malarial regions acquire a natural immunity, that provides protection against malaria, within a few years (Koch 1900). Thus the idea was born that a vaccination against malaria could be developed.

© Springer Fachmedien Wiesbaden GmbH, part of Springer Nature 2023   19
F. Frischknecht, *Malaria*, essentials,
https://doi.org/10.1007/978-3-658-38407-4_4

In 1961, British researchers took blood samples from adults who had suffered from malaria throughout their lives and purified their antibodies. Antibodies are proteins that our immune system produces to defend against infectious agents. They injected the antibodies into children suffering from malaria. Surprisingly, this led to a rapid and sharp decrease in the number of parasites in the blood of the children (Cohen et al. 1961). However, the decrease was not complete and the parasites eventually continued to multiply. This trial, however, was also seen as an indication that a vaccine against malaria could be produced. People 'only' had to be able to produce such antibodies themselves after a vaccination.

Vaccines usually consist of individual proteins of the pathogen or attenuated pathogens. When these are injected into our bodies, our immune system is stimulated and, among other responses also produces antibodies. These antibodies then bind to the pathogens during a subsequent infection and prevent them from functioning; in other words, they protect against infection. For example the antibodies can bind to a virus and block it's entry into a host cell. Or the antibodies cover a bacterium and allow our macrophages to eat the pathogens more efficiently. However, what often works very well with some viruses and has led to the development of many highly efficient vaccines and ultimately even to the eradication of smallpox (Henderson 2009), does not seem to work at all or only very poorly with parasites. This could be because viruses have much fewer proteins than parasites or because parasite proteins occur in too many different variants. The latter is precisely the problem with HIV, the viral cause of AIDS. The HIV virus mutates so quickly that it constantly produces new variants of its few proteins and therefore there is no vaccination available against HIV despite much work and many intelligent attempts. Many scientists see this complexity as a special challenge to work on vaccines against HIV and malaria, in addition to the desire to do a service to humanity. In the following I will discuss the two possible approaches to developing vaccines against malaria as we know it from viruses: The administration of single proteins or the administration of attenuated parasites.

4.2   Proteins as Vaccines

Similar to viruses, malaria parasites use proteins on their surface to infect cells of our body. When a malaria parasite encounters a red blood cell, it must first adhere to it and then penetrate it. While viruses often fuse with the cell's membrane or induce the cell to swallow them, parasites use their own motor to bore into the cell. In order for this motor to work, surface proteins of the parasite must bind to surface proteins of the red blood cell. As we have already seen in Sect. 3.4, the absence of one of these proteins in humans can prevent the infection of *P. vivax*. Such a protein is also present for *P. falciparum*, but it is only absent in very few people, as its absence causes a severe and extremely rare disease. It is therefore obvious to assume that antibodies against these proteins would prevent the parasites from attaching to the red cell. Hence the antibodies should protect against malaria. In fact, many antibodies against different surface proteins do the same, they completely block either adhesion of invasion of the parasite and hence prevent infection of red blood cells. But unfortunately they do so only in the cell culture dish. Why?

In order to be able to understand complex biological processes, scientists need to conduct extremely precise experiments. This is often not possible in complex settings and therefore scientists generally try to develop simplified model systems. These serve to enable the scientists to change specific parameters. The consequences of these changes can then be studied in reasonably exact detail. Biological systems are often influenced by dozens to hundreds of parameters, for example, a disease can develop differently in a woman or a man, in a child or an adult. But even if the investigation is limited to a certain age and sex, factors such as weight, height, physical and mental well-being, the presence of other pathogens and the history of disease can also play important roles. Hence doing experiments in 'natural' settings is often not possible or extremely expensive.

Malaria parasites can be studied in two different, simplified systems: In the cell culture dish or in the mouse. In the cell culture dish, the human pathogens *P. falciparum* and *P. knowlesi* can be investigated (*P. vivax* not yet) and in the mouse, different rodent malaria pathogens can be investigated. Both systems have great advantages and disadvantages. In the cell culture dish, for example, the binding of the infected blood cells to the blood vessels is neglected. This can take place in the mouse, but the mechanism of this binding in mouse parasites is not the same as in *P. falciparum*. Similarly, the immune system is not present in the Petri dish, whereas it reacts slightly differently to the parasites in the mouse than in humans. Conclusions from rodent parasites to human parasites can therefore only be drawn to a limited extent due to these differences.

However, when a certain number of parasites are placed in a cell culture dish containing uninfected red blood cells, or injected into a mouse, the reproduction of these parasites can be precisely monitored. By adding chemical substances or antibodies, it is then possible to test whether the parasites grow less well. In this way, mice can be infected with parasites to test anti-malarial drugs. Antibodies or attenuated pathogens or pathogen components, such as purified proteins, can also be injected to test whether the stimulated immune system of the mouse influences the parasites. High concentrations of these components or antibodies are often used initially. Subsequently, further experiments are carried out to determine how little is still required to achieve a measurable effect. Unfortunately, scientists often fail to critically examine an effect at high concentrations. Mice, for example, can produce far more antibodies than us humans. So if a *Plasmodium* protein in the mouse triggers the production of antibodies and protects the mice against malaria, this does not mean that the injection of the same protein would also inhibit malaria parasites in humans. The same protein might well lead to the production of antibodies (or other reactions) in humans, but these often do not reach the necessary concentration in the blood to provide effective protection against malaria. Unfortunately, this has been shown again and again with proteins from the malaria parasites (Matuschewski 2017).

Even with the first malaria vaccine approved in 2017, only some of the vaccinated people were protected after four vaccinations and only for a short time. This first vaccine consists of a combination of two proteins: One is required by the parasite in the mosquito, in the skin and in the infection of the liver, the other is a viral protein, which generally stimulates the immune system. Vaccination attempts with other proteins, which the parasite needs for its survival in the blood, in other combinations, also failed. This failure is partly due to the fact that too few antibodies are produced, that the immune system cannot produce the antibodies long enough or that they do not protect against different strains of parasite. *Try, fail, try again, fail better:* Since the administration of individual proteins as vaccines failed, different research groups are now investigating whether combinations of these proteins provide better protection.

4.3   Irradiated Parasites

In 1967 scientists in New York conducted an amazing experiment: They injected irradiated parasites from mosquitoes into laboratory mice. The idea was as simple as it was ingenious: The radiation damaged the parasites' genetic material to such an extent that they were unable to reproduce in the liver cells of the mouse (Nussenzweig et al. 1967). The parasites thus entered the liver but were no longer able to move from the liver into the blood. The scientists hoped, contrary to the state of knowledge at the time, that an unnaturally high number of parasites in the liver would activate the immune system. Should this activation last long enough, the mice would have to be protected against new parasites. The researchers tested this hypothesis by reinfecting the immunized and completely healthy mice with infectious parasites after some time. It turned out that not a single mouse contracted malaria: All mice were completely protected by the vaccine.

Soon analogous experiments with similar results were conducted in monkeys and humans. Over the years, however, a complicated picture emerged. It became clear that a large number of parasites had to be injected several times for protection.

It was also shown that in humans full protection is only given against the same strain of the parasite. This means that if one protects against a strain of *P. falciparum* from West Africa, a strain from East Africa can still infect in a largely normal way. And the ability of the parasites to reproduce sexually led to a large number of slightly different strains that manage to trick the immune system.

Nevertheless, a company was founded in the USA with the aim of using the attenuated, irradiated parasites as vaccines in Africa and for travelers. Under the name Sanaria ('good air'), the researchers succeeded in breeding mosquitoes on a large scale and infecting them with parasites. In sterile rooms, the parasites are isolated from the germ-free mosquitoes by hand, cleaned and frozen. The company now dispatches these frozen parasites to the world and many of the experiments carried out in the 1960s and 1970s are currently repeated on a larger scale and in slightly different ways. One of these experimental variations involves the use of infectious parasites instead of irradiated parasites. At the same time as these potentially deadly parasites are injected, a drug is given to kill the parasites as soon as they leave the liver and enter the bloodstream. Scientists in Tübingen, Germany showed that only three injections of 50,000 parasites each provide complete protection against the same strain of *P. falciparum* (Fig. 4.1; Mordmüller et al. 2017). Critiques wonder whether this method can be used in Africa to the extent necessary to vaccinate hundreds of thousands of people. However, initial pilot tests are underway in several African countries.

4.4   Genetically Modified Parasites

As mentioned in Sect. 4.3, vaccination with live parasites presents many potentially insurmountable hurdles and risks. For example, irradiated parasites die early in the liver and therefore only trigger sub-optimal immune responses. The administration of fully infectious parasites requires the administration of a drug, and if it is not taken properly, vaccination could lead to malaria and even death. Scientists therefore began to look for ways to use genetic engineering to create parasites that stop growing in the liver at different times (Matuschewski 2017). These parasites could then be compared and the most suitable ones selected for efficient vaccination. It is hoped that fewer parasites will require less frequent vaccination to produce full protection. It is also possible to produce parasites that carry additional proteins that improve the immune response or provide protection against different strains.

In 2004, researchers in Heidelberg, Germany succeeded for the first time in producing a genetically manipulated parasite that enters but does not leave the liver

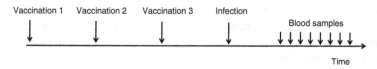

Fig. 4.1 Vaccination scheme – the number of vaccinations depends on the type of vaccination. In the case of experimental vaccination, this is followed by a controlled infection by parasites. The success of the vaccination is examined by taking blood samples and counting the numbers of parasites. (Source: Own representation)

(Mueller et al. 2005). In animal experiments, the mice 'vaccinated' with these parasites were completely protected against a new infection (Fig. 4.1). Ten years later these experiments were repeated with similar results using genetically manipulated *P. falciparum* parasites in humans by researchers in Seattle (Kublin et al. 2017). This delay shows among other things how complex the genetic manipulation of *P. falciparum* is. One proceeds as follows: In both *P. falciparum* and the rodent malaria pathogens, genetic manipulations are performed in the parasite stages that infect red blood cells. For this purpose, a piece of genetic material is produced in a test tube and multiplied in the intestinal bacterium *Escherichia coli*. This piece is constructed in such a way that when added to the parasite, it can be incorporated into its genetic substance, thereby specifically switching off a gene. The added artificial genetic material is introduced into the parasites with the help of an electric field. However, only every 10,000th or 100,000th parasite absorbs the artificial genetic material and makes the desired change. These few parasites must then be separated from the others. The so-called 'selection' is achieved by a drug that kills all those parasites that have not absorbed the artificial genetic material. This in turn is possible by inserting a gene for resistance to the drug into the artificial genetic material (Fig. 4.2).

If a modified parasite is produced, it must first be precisely characterized. To do this, the researchers must answer the following questions: Does the modified parasite grow as well as the unmodified parasites in blood cells? Can it produce gametocytes, that is, the sexual forms of the parasite? Can it infect mosquitoes? If one of these steps fails, then the experiment has failed and one has to start over, that is, switch off another gene. But which of the 5500 genes is the most suitable? How did the researchers from Heidelberg come to 'their' gene?

Like any cell, the parasite needs a number of genes for its survival, but has other genes that are not absolutely necessary. The latter help it to grow better, produce more offspring or move faster. These genes are also important, because without such genes a parasite would eventually lose in competition with another parasite. If

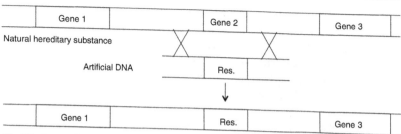

Modified genetic material

Fig. 4.2 Scheme of a simple genetic manipulation – on a natural genetic substance (DNA) there are three different genes in succession. An artificial DNA is produced, in which the areas before and after gene 2 (marked by X) contain a new gene called "Res." (short for resistance gene). If the artificial DNA is introduced into the nucleus of the cell, it can replace gene 2 with the resistance gene, which is located between gene 1 and gene 3 in the modified genetic material. (Source: Own representation)

we now try to switch off a gene that is absolutely necessary for growth in the blood, we rapidly find that this is not possible. The desired parasite is not viable and dies. However, if we succeed in switching off a gene, then we know that the gene is not absolutely necessary, at least not for the survival of the parasite in the blood. But it could be that such a gene is absolutely important for survival in the mosquito or liver.

It is therefore easy to see that a lot of trial and error is required in the search for the best gene. In order to shorten the list of genes a little, different preliminary investigations can be carried out. For example, it is possible to determine the composition of the parasites in so-called mass spectrometers. For the development of these machines the Nobel Prize for Chemistry has already been awarded twice. They make it possible to determine the type and quantity of proteins in a sample which can be a specific type of cell or the different stages of a parasite. Other possibilities include the analysis of active genes, that is, those genes whose information is converted into protein at a certain point in time. So if these analytical methods are carried out at different stages of the parasite's development, the result is a long catalog. In this catalog you can find, for example, a gene 'A001', which is active in the early blood stages, but not in the gametocytes, not in oocysts, but again in the liver. And a gene 'A002', which is only active in gametocytes, and so on. Thus, a list of genes that are only active in the liver can be drawn up and it can be assumed that these are important for the growth of the parasite in the liver. Although none of these lists are perfect and still require a certain amount of trial and error, the lists do limit the trials to a manageable number.

From such a list (and from other lists of other procedures) the scientists selected their genes and generated different parasites, a few of which had the desired characteristics and could enter the liver but could not get out. However, vaccination trials with these parasites showed that they did not produce better immunity than those who were irradiated. One problem was that the genetically modified parasites interrupted their development in the liver very early on. So researchers moved on to test other genes to generate parasites that arrest late (Goswami et al. 2020). Naturally, producing parasites that only stop growing later has the risk that a few do get into the bloodstream and cause an infection. To circumvent this problem, the researchers created parasites that lacked several genes. However, this showed that these parasites are very safe as vaccines, but unfortunately not 'stronger' than the irradiated ones (Kumar et al. 2016). Novel approaches now aim to insert a kind of genetic switch into the parasites, which kills the parasites at a desired stage. There are various approaches to this as well, and we can eagerly wait to see whether these will work better.

Public Health and the Hubris of Modern Science

Abstract

Much has been done over the centuries to combat malaria. Many new medicines have been developed, swamps drained, and insects killed. It is unclear whether modern science or tried and tested methods will mainly contribute to fighting malaria in the future. Clearly, the current efforts will not be enough to eradicate malaria.

Malaria was already known to the advanced civilizations in China and the Middle East thousands of years ago. Chinese travellers to malarial areas were asked to organize a wedding for their wives with another man for the case that they did not return. China is also the source of the most commonly used anti-malarial agent today: Artemisinin, isolated from the mugwort like plant *Artemisia annua*. The genus of the plants was named after the Greek goddess of hunting and forests, Artemis, who was also worshipped as the healer of young girls and mothers. The active ingredient artemisinin was discovered in the 1970s by a Chinese research group led by Youyou Tu, who received the Nobel Prize for Medicine in 2015 (Table 5.1).

In the western world, however, the first remedy was the bark of the cinchona tree. 'Discovered' in South America, the bark was sold to Europe, where a 'quack' saved the life of the Pope. Oliver Cromwell, as a convinced opponent of the Pope, did not want to consume the powder (Table 5.2). For a long time it was the only remedy against malaria (Rocco 2004). Friedrich Schiller, the famous German poet

© Springer Fachmedien Wiesbaden GmbH, part of Springer Nature 2023
F. Frischknecht, *Malaria*, essentials,
https://doi.org/10.1007/978-3-658-38407-4_5

Table 5.1 Nobel Prize winners

1902	Ronald Ross for discovering that mosquitoes transmit malaria
1907	Alphonse Laveran for the discovery of the malaria parasite
1927	Julius Wagner-Jauregg for his treatment of neurosyphilis by infection with malaria parasites, which were subsequently killed with quinine
1948	Paul Hermann Müller for developing the insecticide DDT
2015	Youyou Tu for the discovery of artemisinin as antimalarial agent

Table 5.2 Selection of famous malaria patients

Alexander the Great	Commander, 356–323
Alarich I	Conqueror of Rome, 370–410
Otto II	Holy Roman Emperor from 955 to 983
Henry VI	Holy Roman Emperor from 1191 to 1197
Genghis Khan	General, 1162–1227
Dante Alighieri	Poet, 1265–1321
Christopher Columbus	Discoverer, 1451–1505
Vasco da Gama	Discoverer, 1460(?)–1524
Pope Urban VII	Twelve-day Pope in September 1590
Oliver Cromwell	Lord Protector, 1599–1658
George Washington	First President of the USA, 1732–1799
Friedrich Schiller	Poet, 1759–1805
Abraham Lincoln	President of the USA, 1809–1865
David Livingstone	Missionary, 1813–1873
Ernest Hemingway	Writer, 1899–1961

took it as did George Washington. The seeds of the tree, like those of the gum tree later, were smuggled from South America and cultivated in the tropical colonies of the Europeans worldwide. The plantations were therefore important targets for the Japanese during the Second World War. In 1820, two French scientists isolated quinine from the bark, which could eventually be produced chemically. A chemically related product was the synthetic chloroquine, a miracle cure for malaria, as it killed the parasite more reliably and was incredibly cheap (Table 3.1). Used together with the insecticide DTT, it was believed to allow malaria eradication in the 1950s. However, DDT was so widely used in agriculture that it became an environmental problem and mosquitoes became resistant to it (Carson 2002). At the same time the parasites developed resistance to chloroquine. Moreover, many of the workers involved in the eradication programs realized that they did not have long-term career opportunities and organized themselves into trade unions or looked for new jobs. The program lost the workers who build the basis of the ıda. The belief in the global eradication of malaria, despite the _ɔpe and North America, was lost (Najera et al. 2011). Shortly after ɔuɪlding of the Berlin Wall, three stamps of the German Democratic Republic, communist East Germany, appeared with the inscription "The world united against malaria". United it wasn't. And indeed the eradication of malaria was soon no longer so important to the world. The national and international programs were scaled back with the foreseeable consequences: Hundreds of millions of people fell ill and died of malaria again. In the early 1970s, 'only' about 100 million people a year contracted malaria, but 20 years later it was almost 500 million. It was not until the late twentieth century that a broad alliance of private donors (e.g., the Bill and Melinda Gates Foundation) and public organizations (e.g., the WHO) came together again to fight malaria. From an annual 500 million cases, the number of cases decreased to about 200 million in 20 years. But now this number seems to be stabilizing, despite great efforts, the number of cases has not declined for several years (WHO 2020).

Most scientists and politicians agree that we need new methods, new drugs and finally an effective vaccine to fight malaria. However, we also know that proven methods (use of insecticides, bed nets, drugs – Tables 5.3 and 5.4) can already enable us to displace malaria, as the reported victory of China over malaria in 2017 impressively confirms (Feng et al. 2018). However, this requires above all a functioning health care system, which requires the will of those in power to create and maintain one. Over 30% of malaria-related deaths in recent years occurred in Nigeria and the Congo. While Nigeria is formally a democracy, it is one of the most corrupt countries in the world and so the high revenues from oil production disappear into the pockets of those in power. In recent years, the Congo has experienced

Table 5.3 Selected methods of malaria control

Correct diagnosis with specific treatment
Educating the population on the causes of malaria (and other diseases)
Sleeping under (insecticide-treated) bed nets
Attaching mosquito screens to windows
Avoidance of staying outside during twilight
Spraying insecticides in houses
Prevention of open water reservoirs
Drainage of swamps and open water reservoirs
Control of mosquito larvae
Preventive measures during travel
Taking medication for prevention and treatment

Table 5.4 Figures on malaria control (WHO 2020)

90% of the more than 200 million annual malaria infections occur in Africa
In recent years, the world spent annually US$ 2.5–3 billion on malaria control
800 million impregnated bed nets were distributed in Africa between 2011 and 2016
In just 2 years, 700 million artemisinin combination therapies were administered
The number of malaria cases in politically unstable Venezuela increased from under 50,000 to over 240,000 between 2011 and 2016

the deadliest warlike conflict since World War II and is even further away than Nigeria from sharing the wealth of its natural resources with its people.

If the world were to succeed in displacing malaria with existing methods except in a few countries in Central Africa, the malaria parasites could spread again from there. The population in the neighboring countries would again fall ill due to the lost immunity, just like small children who get malaria for the first time. In fact, it is a perfidious characteristic of malaria that natural protection is only acquired through several infections and that this protection is lost again when no infection has taken place for a long time. The goal must therefore be to eradicate the disease worldwide, otherwise epidemics will continue to occur. But how can this be done with the political instability mentioned above? A cheap malaria vaccine, which can be given once, seems to be indispensable. But it does not exist and will probably not exist in the next 30 years, that much can unfortunately already be predicted at the present state of research, despite all progress (Desowitz 1993; Matuschewski 2017).

Another way would be to eradicate the *Anopheles* mosquitoes. This seems to be possible by means of new technical methods that are promising in laboratory experiments (Kyrou et al. 2018). But will there be a political consensus if this requires the release of billions of genetically modified mosquitoes? Another strategy would be to modify a parasite so that it still infects people, but they do not fall ill. Could such a "vaccine parasite" be improved so that it displaces the natural parasite? This also raises complex ethical questions, as humans would be infected with genetically modified parasites. If the acceptance of the application of modern genetic engineering is a generation or two away, there is probably little hope of using modern science to defeat malaria in the near future. Unless a chance discovery allows an unexpected leap. While this miracle cure is being sought, efforts to combat malaria by conventional methods must continue tirelessly at the public health and political levels. According to WHO estimates, the funds made available for this would have to be more than doubled, from US$2.7 billion per year in 2016 to approximately US$6.5 billion. This corresponds approximately to the annual costs caused by malaria in Africa alone (WHO 2020).

What You Can Take from This *essential*

- Malaria is a disease of poverty that is currently restricted to the tropics, but was once widespread in Europe and North America.
- Malaria is caused by various parasites and transmitted by *Anopheles* mosquitoes.
- Malaria parasites multiply in different host cells in humans (in the liver and blood) and in cysts on the stomach of mosquitoes.
- Malaria parasites differ greatly from one another, which makes the development of an efficient vaccine difficult.
- The great health pressure of malaria has altered the human genome over the course of evolution, leading to a number of new diseases such as sickle cell anemia.
- Malaria parasites also attack birds and mice, among others. These parasites can be used to gain new insights in laboratory experiments.
- The principles of genetic modification of parasites to produce attenuated parasites as live vaccines.
- Different methods of malaria control.

© Springer Fachmedien Wiesbaden GmbH, part of Springer Nature 2023
F. Frischknecht, *Malaria*, essentials,
https://doi.org/10.1007/978-3-658-38407-4

Glossary

Bacteria *Class* of organisms (prokaryotes) that can survive on their own. Compared to human cells, they have a simpler structure. Only very few types of bacteria are also pathogens of diseases such as diarrhea and tuberculosis.

Beta-globin *Protein* in red blood cells that forms a component of hemoglobin. Beta-globin is mutated in people with sickle cell disease.

Bone marrow *Central* part of some bones, which is important, among others, for the formation of red blood cells.

Chloroplast *An* organelle of plant cells and also of certain unicellular organisms that converts light into energy. The chlorophyll in chloroplasts gives leaves their green color.

Chromosome *Parts of* the genetic material on which the genes are strung. Humans have 46 chromosomes. 22 "normal" and 2 "sexual" (X and Y). Women have two X chromosomes and men each have one X and one Y chromosome. Malaria parasites have 14 chromosomes.

Eradication of smallpox *In 1977,* the last person was naturally infected with smallpox, after 300 million people died of it in the twentieth century. Following a major vaccination campaign, the World Health Organization declared small-pox eradicated in 1980.

Gametocytes *In* Plasmodium: The forms of the parasite that are taken up by the mosquito. They appear in male or female form.

Gene *Part of* the genetic material that contains the information for the production of one or more versions of a protein.

© Springer Fachmedien Wiesbaden GmbH, part of Springer Nature 2023
F. Frischknecht, *Malaria*, essentials,
https://doi.org/10.1007/978-3-658-38407-4

Hemoglobin *Red* blood pigment responsible for the transport of oxygen from the lungs to the tissues and the removal of carbon dioxide from the tissues to the lungs.

Lymph *Fluid* that is removed from tissues in so-called lymph vessels.

Mitochondria *Energy* producing organelles in eukaryotic cells.

Mutation *Change* in the genetic material. A mutation can lead to a change in the structure and function of a protein, among other things.

Oocyst *Parasitic* cyst form of Plasmodium, which is formed on the stomach wall by ookinetes and in which sporozoites form.

Ookinete *Mobile* form of Plasmodium that penetrates the stomach wall of the mosquito and forms an oocyst there.

Organelle *Component of* a eukaryotic cell that performs certain tasks, such as the energy-producing mitochondrion or the nucleus that stores genetic material.

Parasite *Class* of pathogenic organisms (eukaryotes) that can contain cells of similar complexity to human cells. Some parasites consist of a single cell, while others can weigh many kilograms.

Sickle cell disease Caused by a mutation in beta-globin. If the mutation was inherited only from the father or mother, it protects against severe malaria. However, if the mutation is inherited from both parents, a severe, fatal disease results.

Sporozoite *Mobile* form of Plasmodium that is formed in oocysts and transmitted by the mosquito.

Viagra *Drug* used to treat erectile dysfunction in men.

Viruses *Are* neither alive nor dead, as they can only multiply within a host cell. Only a few types of viruses can cause diseases such as diarrhea, AIDS, flu, and herpes.

WHO *World Health Organization*, see www.who.int.

Nájera JA, González-Silva M, Alonso PL (2011) Some lessons for the future from the Global Malaria Eradication Programme (1955–1969). PLoS Med 8:e1000412. https://doi.org/10.1371/journal.pmed.1000412

Nussenzweig RS, Vanderberg J, Most H, Orton C (1967) Protective immunity produced by the injection of x-irradiated sporozoites of *Plasmodium berghei*. Nature 216:160–162

Ramdani G, Naissant B, Thompson E, Breil F, Lorthiois A, Dupuy F, Cummings R, Duffier Y, Corbett Y, Mercereau-Puijalon O, Vernick K, Taramelli D, Baker DA, Langsley G, Lavazec C (2015) cAMP-Signalling regulates gametocyte-infected erythrocyte deformability required for malaria parasite transmission. PLoS Pathog 11:e1004815. https://doi.org/10.1371/journal.ppat.1004815

Rocco F (2004) The miraculous fever-tree, the cure that changed the world. Harpercollins, New York

Shortt HE, Garnham PC, Malamos B (1948) The pre-erythrocytic stage of mammalian malaria. Br Med J 1:192–194

Sturm A, Amino R, van de Sand C, Regen T, Retzlaff S, Rennenberg A, Krueger A, Pollok JM, Menard R, Heussler VT (2006) Manipulation of host hepatocytes by the malaria parasite for delivery into liver sinusoids. Science 313:1287–1290. https://doi.org/10.1126/science.1129720

WHO (2020) World malaria report. https://www.who.int/teams/global-malaria-programme/reports/world-malaria-report-2020. Accessed 16 Dec 2020

Wirth D, Alonso P (2017) Malaria: biology in the era of eradication. Cold Spring Harbour Laboratory Press, Cold Spring Harbour (Freely available on the publishers site)

CPSIA information can be obtained
at www.ICGtesting.com
Printed in the USA
LVHW041934180323
741943LV00001B/76